GLOBAL WARMING FOR DIM WITS

A SCIENTIST'S
perspective of
CLIMATE CHANGE

James R. Barrante, Ph.D.

Universal-Publishers
Boca Raton

Global Warming for Dim Wits:
A Scientist's Perspective of Climate Change

Universal-Publishers
Boca Raton, Florida • USA
2010

ISBN-10: 1-59942-861-X
ISBN-13: 978-1-59942-861-1

www.universal-publishers.com

Library of Congress Cataloging-in-Publication Data
Barrante, James R., 1938-
Global warming for dim wits : a scientist's perspective of
climate change / James R. Barrante.
   p. cm.
Includes bibliographical references.
ISBN-13: 978-1-59942-861-1 (pbk. : alk. paper)
ISBN-10: 1-59942-861-X (pbk. : alk. paper)
1. Climatic changes. 2. Climatic changes--Environmental
aspects. 3. Greenhouse gases. 4. Greenhouse gases--
Environmental aspects. I. Title.
QC903.B37 2010
551.6--dc22
                                        2010002946

CONTENTS

Preface   vii

1: *I Dare You To Teach Me*   1

2: *Is the Globe Actually Warming?*   13

3: *That Notorious Pollutant Carbon Dioxide*   31

4: *How Can We Study the Greenhouse Gas Effect?*   43

5: *It's The Oceans, Dim Wits*   57

6: *When Is a Greenhouse Gas Not a Greenhouse?*   71

7: *Filling In The Blanks*   89

Source Material/Suggested Reading   111

List of Figures   113

Glossary   115

PREFACE

When the globe came out of the last ice age 20,000 years ago, do you think the earth's inhabitants were aware of it? Do you believe that they could tell that the earth was warming? Neanderthal man by this time was extinct. Homo sapiens thrived in an extremely cold European climate. The Americas were beginning to be populated as low sea levels and a frozen Bering Sea formed a land bridge from Asia to Alaska.

Today, there seems to be a lot of concern by the earth's population about global warming and climate change. Governments appear to be on the brink of spending trillions of dollars to try to stop it. It took 10,000 years to come out of the last ice age. If we know anything about climate change, we know that it changes very slowly. Human beings live for about 70 years. Back 20,000 years ago, human life span was most probably not that long, maybe 40 or 50 years. Certainly, ten generations would have been less than 1000 years. So, let me ask the question again. Do you think Homo sapiens knew that the climate was changing 20,000 years ago, when they went about their daily routine and lived for, say, 50 years or so? Is it probable that a grandfather said to his grandchild, "You know, when I was little, it was a lot colder around here"?

Perhaps a better question is: do you think it is possible to detect any kind of climate change by studying global temperature for 150 years? Considering that it takes about 10,000 years for the earth's climate to change, it almost seems like a silly question, like asking whether it would be scientifically valid to study ocean tides for a two-hour period? I guess the next question would be: do you think it makes any sense what so ever to spend one dime trying to stop the climate from changing?

In the past 400,000 years the climate of the earth has changed only four times. All scientific data point to the fact that this change is controlled externally, occurring approximately every 100,000 years. During one of these cycles, the earth spends about 10,000 years or less in a warm period, such as the period we are experiencing today. Then the climate slowly drops back to ice age and remains there for about 80,000 years. Finally, very rapidly over a 10,000-year period (10,000 years is rapid on a global time scale) the globe returns to the warm period. Scientists have suggested that this cycling of going in and out of ice age every 100,000 years that we know we have been in for at least the last 400,000 years has occurred only twice in Earth's known history – now, and a period 300 million years ago. These are the only two periods in the last 600 million years when atmospheric CO_2 levels were less than 400 ppm and global temperatures were the same as they are today. During the whole 100,000-year cycle, the temperature of the globe continuously wiggles up and down about two degrees Fahrenheit over a period lasting about 200 to 300 years. For the past 150 years, we have been experiencing one of these up-wiggles.

This book was written to convince you that this wiggling of temperature of about 2 or 3 degrees Fahrenheit has nothing to do with climate change. What causes this temperature wiggle is not known? It may simply be static (what scientists call noise). Certainly, greenhouses gases may have some part in the process, but they are not a major player. We know this because the mathematical relationships connecting global temperature to greenhouse gas levels, if greenhouse gases were a major player, are simply not there. We also know that the inhabitants of the planet do not cause it and cannot stop it, as evidenced by the fact that it has been going on for at least 400,000 years. Moreover, it is not possible for creatures of 70 years or so to detect climate change, unless some catastrophic event such as being hit with a comet or asteroid took place. Trying to perceive events that occur over a 10,000-year period is not within the realm of our experience.

The fact that global temperature has risen 1.4°F or so in the last 150 years, or the fact that global temperature has not increased since 1998 proves only two things scientifically: that global temperature has risen 1.4°F or so in the last 150 years and that global temperature has not increased since 1998. It is not evidence that the temperature of the globe will continue to rise, fall, or go side-ways. And it certainly is not proof that the climate is changing. If any human being produces evidence that the climate is presently changing, that evidence we know is not related to climate change. We can only determine climate change by looking at the past. It's like watching a tree grow. We cannot see it happening. We only know it has happened by observing that the tree is taller at some later date. Who knows? The Earth may have started its trek back to ice age. Perhaps in 10 generations, humans will notice the climate change by looking back in time.

This book is not a scientific textbook. It was written in plain language for the layperson that wishes to learn about the science of climate change. While the language does get a little technical in some sections, every attempt was made to keep the science simple and understandable.

Anyone who needs further information about a term or concept is urged to refer to the *Glossary*.

One of the most upsetting aspects of the global warming debacle is the misinformation being given to our children by individuals who should know better. I am hoping that this book becomes a popular read for children. Believe me, they are extremely intelligent and will understand it. Moreover, it should convince them that, contrary to popular belief, the oceans will not rise and flood the land; the polar bears will survive just fine; carbon dioxide is no more a pollutant than is water – plants need it to survive; we should stop burning fossil fuels, not because they produce CO_2, but because they are precious and momentary; we should protect the environment, not control it, because we are part of it; climate change is a natural event that human beings cannot control; and most importantly that they are safe and that the sky is not falling .

No one writes a book in a vacuum. I have been working in this area of science for a number of years, and there are a number of people that I must thank. First, I would like to thank Professor Gregory Kowalczyk, an Environmental Analytical Chemist and colleague of mine at Southern Connecticut State University, for allowing me to bounce my ideas off him. Our many discussions and his valuable input are major reasons why I wrote this book.

I also would like to thank some individuals who reviewed the manuscript and made valuable suggestions: Mr. David Whalen, Mrs. Sharon Barrante Adkins, and Ms. Kimberly Barrante. A special note of thanks is due my son, Mr. Stephen Barrante, for his valuable suggestions concerning the layout of the book and for his beautiful cover design. To Jeff Young, Christie Mayer, and the people at BrownWalkerPress/Universal Publishers, thanks for taking on this project and producing a quality product. And finally, I would like to thank my wife, Marlene, who has put up with me during the process of writing the book and for her thoughtful input. I could not have done this without her.

I take full responsibility for any errors found in the book. If you stumble on any, I would appreciate your pointing them out to me at jim@tenwheeler.net.

James R. Barrante
Cheshire, CT
January, 2010

1

I DARE YOU TO TEACH ME

When asked to describe the chemical properties of water on a basic chemistry exam, a student in my class once answered, "Water is made up of two elements, H and O." Then, with a stroke of genius, the young man went on to add, "To make holy water, you have to take regular water and boil the H out of it."

It is estimated that the human brain consumes about 10 grams of the sugar glucose in an hour. It "burns" this sugar to produce carbon dioxide and water and energy. If one calculates the power output of this process, one will find that it is equivalent to about 40 watts. That is, the human brain is about as "bright" as a 40-watt light bulb. We humans, all of us, are definitely "dim wits" – a lot of heat and not much light. Frankly, it is amazing that the human race has survived as long as it has.

This book was written for dim wits. The dictionary defines a dimwit as a stupid or silly person. That definition doesn't apply in this case. As a physical scientist and somewhat knowledgeable in the physical chemistry of the atmosphere, I am going to re-define a "dim wit" as someone who believes that greenhouse gases, and in particular carbon dioxide, could actually control the climate. Such individuals generally fabricate their version of science to fit their own agenda. When it comes to dim wits, there is a lot of truth to the old saying, "in one ear and out the other!"

I found after being a university chemistry professor for over 40 years that it is nearly impossible to educate individuals who do not desire to be educated. Dim wits, for some particular reason, hold on to a science that has not been tested experimentally and in many cases is known to be wrong. Unfortunately, many of our politicians fall into this class, and this is dangerous, since they have the power to control our lives.

"Good" science is based on the scientific method. The scientific method is very simple to understand. One first proposes a theory. Then one takes the theory into the laboratory and tests it by experimentation to see if the experiments give outcomes that are consistent with the theory. If they do not, then the theory is not valid. It's as simple as that. One then can either throw out the theory as being bogus or modify the theory and try again. As the famous American physicist Richard Feynman once said about scientific theory, and I'm paraphrasing, "It doesn't depend on how many people believe it, who believes it, how famous the person is who proposes it, or how clever or correct the theory sounds. If it doesn't hold up to scientific experimentation, it's false."

A computer model is not based on the scientific method, unless it can be tested experimentally – a very difficult thing to do. This is because computer models generally contain a number of "if-then" statements – if this happens, then that will happen. Also, we must keep in mind that a computer is no more intelligent than the person who programs it, and that limits its computing power to 40 watts. For example, most of the predictions about the consequences of global warming made at Kyoto have not happened, and many have been found to be wrong. In fact, a number of those scientists involved in the original work at Kyoto are now having second thoughts about their work. Moreover, not one computer model correctly predicted that the globe would stop warming in the late 90's. How could it? It would be like asking a computer to predict the exact date and time a particular leaf will fall from a tree, or who will find the candy bar with the golden ticket. It simply cannot be done with any precision.

And just to set the record straight: when it comes to climate change, there is no more scientific significance in the fact that global temperature has not changed in the last ten years, than in the fact that global temperature has increased a degree in the last 100 years. It is not well publicized, perhaps intentionally, that in the last 100,000 years global temperature has risen as least $0.8°C$ in a 150-year time period thousands of times. So there is nothing unusual or different about the increase in global temperature from the mid-1800s to the present.

One of the most difficult things with which a scientist is faced when he or she studies the universe, is to correctly define the boundaries around that portion of the universe that he or she intends to study. Even the best-trained scientists have trouble with this. Some areas of science refer to this as distinguishing the signal from the noise. Incorrect boundaries around systems can lead to erroneous interpretations of experimental results, or even to performing the wrong experiments in the first place.

A good example of this is the idea that hot water freezes faster than cold water. If you do the experiment with identical amounts of water, one at, say, $90°F$ and the other at $50°F$, both under the same cooling conditions, you will find that sometimes the hotter water will freeze before the colder water. This has become known as the *Mpemba Effect*. How is this possible? It sounds counter intuitive. The simple answer is that the boundaries of the systems are different. One is assuming erroneously that the only difference between the hot and cold water is their average temperature. Not so! Hot and cold water are much different from each other – things like evaporation effects, convection effects, and surface effects must also be included, along with temperature. So actually, the effect is a comparison of apples and oranges.

Consider another example: Is it valid to make predictions about future climate changes based on the behavior of the climate over a 200-year period? Is there any difference between looking at global temperature for a 200-year period and looking at global temperatures for a week or an hour? If I were to tell you that the fact that it is warmer today than it was yesterday is proof that the climate is changing, you hopefully would say that the observation is ridiculous. Daily temperature changes are not proof of climate change. If I were to tell you that the fact that it is warmer in July in New York City than it is in December is proof of global warming, again I hope you would say that the observation is insane. Seasonal temperature changes are not proof of climate change. How do we know this? We know this because it is not consistent with our experience. If I were to say that it is warmer today on the average than it was 150 years ago, is that proof of climate change? I bet that many people would say that it is.

But why is this any different from the previous statements? Could it be because we are creatures of 70 years or so (if we are lucky) and 150 years seems to us like a long period of time? To a tree, 150 years is more like 20 or 30 of our years. To the earth, 150 years is a blink. On a global time scale, there is no measurable difference between 150 years and 1 minute. It would be same as asking humans to distinguish between a thousandth of a second and a millionth of a second. A period of 200 years is close to the error of time measurement. How do we know this? We know this because, again, it is not consistent with our experience.

If I were to ask you to choose a reasonable time period between time-temperature data points in which to study global climate change, what would you choose? Would measuring temperature once every week give you enough information? How about one measurement every year? Most individuals would probably agree that we shouldn't go longer than a year to gather information. But most people, particularly our politicians, are not trained scientists.

The scientists who originally collected the temperature data going back thousands of years decided that one data point every 80 to 200 years would be scientifically valid. If they didn't, being good scientists, they would not have published their results. In the 150-year period of the "inconvenient truth" there are only one or two data points. It is quite clear that the scientists who collected the data decided that to look at temperature data closer than every 80 to 200 years would not add anything scientifically valid to the study.

The only way we can understand this idea of climate change is to expand the boundaries of our system. We must look at global temperature over hundreds of thousands of years to see evidence of climate change. When we do, we see something very interesting about global temperature. It does not change haphazardly. It changes in a very orderly, structured way. Moreover, when we look at it over very long periods of time, the changes appear to be periodic; that is, the changes occur in a cyclic pattern lasting about 100,000 years. Now, greenhouse gases may be causing these temperature changes. But the major greenhouse gas on our planet is water vapor, and there is no apparent change in water vapor levels preceding global temperature changes. Ah, it's the CO_2, you say. But if it is global CO_2, what causes the CO_2 to change? Certainly, there were not a lot of people around burning fossil fuels or driving SUV's 400,000 years ago. Some people will argue, "The changes today are different from what they were 400,000 years ago!" But that is not what the data shows. In fact, they are exactly the same. If there is any difference, it is that the planet has been warm too long and should be heading back to ice age.

In subsequent chapters we will explain the accepted, scientifically consistent behavior of the global temperature and global climate change. It is not based on any political agenda, except, perhaps, to help dispel the erroneous notion that any species on this planet can actually affect the climate of the globe. The level of carbon dioxide in the atmosphere today is about 0.036%. In a crowd of 10,000 people, this is about four people. It is the intention of the nations of the world (including the United States) to cut greenhouse gas emissions in half by 2050. Greenhouse gas emissions, however, are only a small portion of this 0.036%. Cutting greenhouse gas emissions in half will lower the percentage of CO_2 down to maybe 0.03%? That is the same as removing one person out of those 10,000 people that I mentioned above. It is extremely difficult to believe that this small change in the CO_2 levels in the atmosphere will have any profound effect on the climate of the globe. Yet, many dim wits believe this.

The science of global warming is no different from the science of any other thing. It should not be centered on any political or economic agenda. It is neither liberal nor conservative, right wing nor left wing, capitalistic nor socialistic. And most importantly, science should never be centered on dogma. That is the realm of religion.

2

IS THE GLOBE
ACTUALLY WARMING?

"The relationship (between atmospheric CO_2 and global temperature) is very complicated. But there is one relationship that is far more powerful than all the others, and it is this. When there is more carbon dioxide, the temperature gets warmer." This statement of former Vice President Al Gore in "An Inconvenient Truth" is perhaps one of the most far-reaching statements of recent times that will have a profound effect on the future economies of the world.   Mr. Gore based this statement on, among other things, examining graphs of global temperature and atmospheric CO_2 levels for at least the last 200 or so years. If we look at the behavior of atmospheric CO_2 and global temperature over the past few hundred thousand years, we would see a striking correlation between CO_2 levels and global temperature. They appear to follow each other very closely, so perhaps Mr. Gore's observation is correct.

How were global temperatures measured hundreds of thousands of years ago? Obviously, accurate thermometers have only been around for a couple of hundred years. To answer that question, we need to learn a little basic chemistry. Oh, it's not that hard! As the student described in Chapter 1, water is a substance made up of two elements hydrogen and oxygen. All atoms of an element, the smallest, chemically indivisible particles of that element, however, are not the same. Some atoms of an element are heavier than others because the nucleus, the core of the atom, contains some extra benign particles called *neutrons*. Neutrons basically do nothing more than to change the weight (mass) of the atom. These different types of atoms of a particular element having different weights are called *isotopes*. The majority of all the hydrogen atoms in water are "light" hydrogen; let us call it hydrogen-1. The nucleus of light hydrogen atoms contains no neutrons. A very tiny percentage of hydrogen atoms in water, however, are a heavier form of hydrogen, call it hydrogen-2. This heavy form of hydrogen is called *deuterium* (it's still hydrogen, but it has its own name, because it is an important isotope), and the water that forms with these heavier hydrogen atoms is called *heavy water*. The nucleus of an atom of heavy hydrogen contains one neutron, and this doubles the weight of the atom. The properties of heavy water are quite different from normal (light) water. For example, heavy water is very poisonous to us. Luckily, natural water contains only a minute amount of heavy water, not enough to harm us.

The ratio of heavy water to light water in natural water apparently changes by a very small, but measurable, amount with global temperature. Consequently, if one examines Antarctic ice formed thousands of years ago and

measures the ratio of heavy water to light water, one sup-
posedly can determine the average temperature of the
globe at the time the ice was formed.

A number of years ago, scientists, working at Vostok
Station, Antarctica, took core samples of ice; that is, they
pushed long tubes into the ice and pulled out samples of
ice going back hundreds of thousands of years. The ice
cores were then divided into periods of time, similar to
the rings on a tree stump, and by analyzing the cores, the
average global temperature over these time periods was
determined. The results of these analyses are shown in
Figure 2-1 for the last 400,000 years. It is clear from the
Vostok ice-core samples that global temperature changes
in a very controlled, periodic way. The repetitive structure and shapes of the peaks occurring every 100,000
years or so is quite striking. In fact, notice the doublet
pairs of peaks that repeat just after each large drop in
temperature. That certainly is not a random event.
Moreover, when you analyze the actual temperature
changes taking place, you find that in the 400,000 years

illustrated on the graph, global temperature has increased at least 0.8°C (1.4°F) over a 150-year period thousands and thousands of times. So, it is apparent that there is nothing unusual about the increase of 1.4°F from the mid-1800s to the present. It is simply what the globe does over and over again.

The temperature values on the graph are temperatures relative to present temperature (2000 AD, when the graphs were prepared). We find that when the global temp-

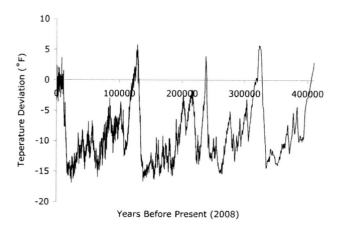

Figure 2-1. Average global temperature versus time for past 400,000 years, calculated from Vostok ice-core data (7).

erature drops down to 16°F below present temperature, the planet experiences a temperature period that we refer to as "ice age." At these very low temperatures, the ocean between Alaska and Siberia freezes, and much of northern North America is under a mile or so of glacier ice. What is striking and surprising about the ice-core data is the indication of the amount of time the globe spends at these ice-age temperatures. Over the past 400,000 years, the globe has spent about 300,000 or so years at temperatures that would be very un-hospitable for North Americans. It appears that these low temperatures are, in fact, the "normal" temperatures of the globe, while present day temperature periods are short lived, lasting for only twenty to thirty thousand years, an extremely short period of time on a global time scale. The globe apparently "likes" to be at ice-age temperatures. So when the former vice-president Al Gore once said that "the globe has a fever," he wasn't kidding. The globe gets this fever every 100,000 years and it lasts for about 20,000 years, apparently whether humans are around or not.

What causes the temperature of the globe to go up and down in the fashion we see on the graph? The simple answer is that nobody has a clue! There are many theories kicking around, but, so far, none of them have been proven to be complete. A safe guess is that it has something to do with the sun. You be the judge. Your guess would be as good as anybody's. One thing we do know, and it is quite clear in Figure 2-1, these changes in global temperature are not random. They are controlled by some event or events in nature that occur *periodically*, which highly suggests that they are not caused by anything occurring on the planet, such as abnormal changes in greenhouse gases, volcanic activity, or the activities of any of the inhabitants of the planet. These would be random events.

So, a part of the question, "is global warming real" has been answered by Vostok ice-core data. The average temperature of the globe is indeed constantly changing, and every hundred thousand years, the globe warms up for a period of about 20,000 years and then goes back to ice age. From the Vostok graphs, it looks like we should be heading back to ice age very soon. This brings up another important question: what is "soon" on a global time scale? Is it 50 years, or, maybe, 100 or 200 years? If you look at the 400,000-year graph, you will see that a 200-year period is about the width of the ink line. It is immeasurable! "Soon," on a global time scale could be as little as 1000 to 2000 years. It took thousands of years to get out of the previous ice age, and even longer to go into these temperature periods. So, to look at temperature change over a 200-year time period and draw some scientific conclusion from these observations is scientifically moronic. The climate does not change in any measurable way over such a short period. Oh, you can take temperature readings over a 200-year time period and see changes, but they have about as much meaning as does taking temperature readings daily and trying to predict climate change from that data.

There is a legend from India that describes 6 blind men trying to identify an elephant that has wandered into their village. The story goes something like this: The first blind man came up from behind the elephant and encountered its tail. "I think," he said, "that the elephant is most certainly like a rope."

The second blind man stumbled into a leg. "What are you saying? It is clear that the elephant is absolutely sturdy like a tree."

The third blind man encountered the ear of the elephant and proclaimed, "I am wanting to tell you that you both are wrong. The elephant is like a giant fan."

The fourth, upon feeling a tusk, yelled out to the others, "What, are all of you crazy? The elephant is like a spear, sharp and pointed."

The fifth blind man, approaching from the front, felt the trunk. "You all are wrong. The elephant is like the heavy branch of a tree."

Finally, the sixth blind man walked into the side of the elephant and proclaimed, "I think that you have all been smoking something. I am standing next to the elephant and it is a wall."

Not one of the blind men got the description of the elephant correct, even though each one correctly identified his portion of the elephant. This is what happens in science, when scientists fail to define the boundaries of the systems they are studying correctly. To get a better feel for how this all works with global temperature, consider the next set of graphs, which takes the 400,000-year data and breaks it down into smaller periods. In other words, what would be our impressions about global climate change, if we were to look at the behavior of temperature over longer and longer periods of time? For example, let's first look at global temperature data for the last 150 years or so. The graph shown in Figure 2-2 is the so-called "industrial revolution" graph cited by Mr. Gore and others. It is a favorite of the IPCC, the International Panel on Climate Change. More data has been included toward recent time periods because more reliable temperature measurements could be made. This gives the graph a more "wiggly" appearance. The graph certainly supports the arguments that present global temperature is

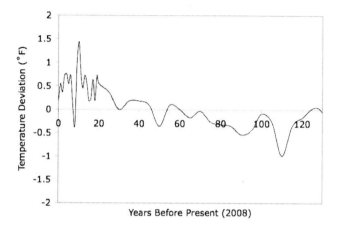

Figure 2-2. Average global temperature for the past
150 years, calculated from Vostok ice-core data (7).

higher now than it has been in the last 130 years. There is
an obvious trend upward. According to the information
presented on the graph, the globe is definitely warming.

But where is the experimental evidence that this
upward trend is caused by increased levels of CO_2 in the
atmosphere. The fact that atmospheric CO_2 also increased during this time period is not scientific proof. It
is coincidence! Even if it sounds like a good idea, and
thousands of scientists the world over believe it to be true,
until it has been tested experimentally and found to be
the case, as Feynman said, it is not valid. And if it cannot

be tested experimentally, then it is certainly not settled science. Any reputable scientist would know this. In a 200-year time period some things are bound to coincide. This period of time also happens to correlate well with increased human life span, the depletion of the ozone layer, numerous wars, and a garbage dump in the Pacific Ocean twice the size of the State of Texas. Perhaps these are causing global temperature to increase.

The next graph in Figure 2-3 shows global temperature over a 300-year period. There is no question from this graph that over the last 300 years global temperature

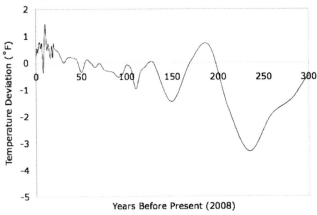

Figure 2-3. Average global temperature relative to the year 2000 for past 300 years, calculated from Vostok ice-core data (7).

has been increasing. One obvious feature of these types of graphs is the amount of temperature wiggle over shorter time periods. One has to wonder if they are real. Moreover, the general smoothing out of the graph is due to the fact that the temperature data are taken from Vostok ice core samples, and the sampling period, that is, the difference between data points, is about 80 years.

The next graph in Figure 2-4 shows global temperature over a 1000-year period. If you believe the data in this graph, you would have to agree that recent

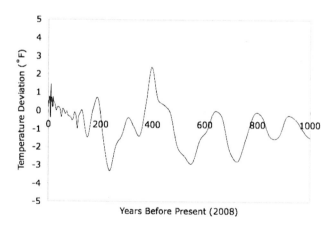

Figure 2-4. Average global temperature relative to the year 2000 for the past 1000 years, calculated from Vostok ice-core data (7).

statements of some suggesting that global temperature in the past 10 years is higher than it has ever been before in human history are erroneous. It looks like 400 years ago, the temperature shot up to about 2.5°F higher than "normal," about a degree higher than the 1.5°F temperature increase of the early 2000's.

Figure 2-5 illustrates the global temperature behavior over the last 5000 years. The "industrial revolution" part of the graph looks rather normal. Also, it is apparent that during this time period, temperatures 1°F or greater above "normal" were not unusual. Moreover, it is clear

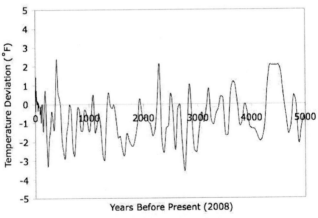

Figure 2-5. Average global temperature relative to the year 2000 for the past 5000 years, calculated from Vostok ice-core data (7).

when you examine global temperature for as long a period as 5000 years that, except for short spurts above and below the zero reference, global temperature has not changed in a permanent way right up to the present. Perhaps the sky is not falling after all.

We can see from this graph that the "industrial revolution" part of the graph, while still present, is beginning to blend into the "noise." In these cases, we will define "noise" as simply the wiggling up and down of global temperature over short periods (200 years) of time, something like static on the radio. We have been experiencing an up-wiggle for the last hundred or so years, a seemingly perfectly natural event, based on the behavior of global temperatures of the past. No one seems to want to address the issue as to whether these up and down wiggles in temperature are real. Do humans have anything to do with this up and down wiggle? There is no scientific evidence so far to suggest that we do. The industrial revolution graph is part of an up wiggle. Of course, like studying the tides for only a three-hour period, if you only study global temperature for the past 150 years, you will never observe this temperature wiggle, and may get the impression that the Earth is warming uncontrollably.

Figure 2-6 describes global temperature over a 20,000-year period. We now can see why it is important to study global temperature over long (to humans) periods of time. From about 18,000 years ago to about

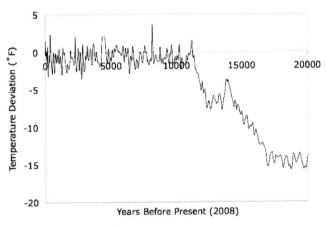

Figure 2-6. Average global temperature relative to year 2000 for the last 20,000 years, calculated from Vostok ice-core data (7). Note that between 18,000 years ago and 12,000 years ago the globe underwent a massive warming of 16°F from its "normal" ice-age temperature to the present day temperature. What caused this massive warming event?

11,000 years ago global temperature increased about 16°F. Now this is climate change! During this time period, the globe came out of the last ice age. The ice bridge between Alaska and Siberia melted. The glacier ice covering most of North America and Northern Europe receded, and polar bears had to move north. Notice that during this transition the wiggling up and down of temperature did not stop, and, in fact, is really beginning to look more and more like noise. But it's too soon to tell. Also, note that the present up-wiggle that we are experiencing today is not remarkable, and is barely noticeable compared to the temperature spike of 5-degrees (from -1°F to + 4°F) occurring about 8000 years ago.

So, going back to Figure 2-1, there is no question that global temperature is constantly changing up and down in a controlled way. This real periodic change in temperature is not observable, unless it is studied over time periods that are at least 20,000 years long. What caused that large swing in temperature from the earth's "normal" temperature? We pompous humans believe that the global temperatures we are experiencing today are the earth's normal temperature. Look at Figure 2-1 and decide what you believe is the normal temperature of the globe. I think you will agree that the normal temperature of our planet is ice age.

Are the short 200-year wiggles that we are presently experiencing important and should they be of any concern? They undoubtedly have been around for a very long time. Could we humans do anything about them? Do you think, perhaps, that greenhouse gas emissions cause them? Apparently, the temperature of the globe is bound to change, whether we are here or not. To believe that we humans could actually control the climate is not consistent with any reasonable interpretation of any global temperature data presented to date.

3

THAT NOTORIOUS POLLUTANT CARBON DIOXIDE

When the earth formed, it had no atmosphere. Any hydrogen or helium gas present was lost to space, since the earth's gravity is too small to hold on to these light gases. Eventually, an atmosphere of approximately 95% water vapor and carbon dioxide formed, due to the out-gassing of the planet's interior (for example, volcanic eruptions). There was no oxygen gas present. Oxygen gas is one of the most poisonous, reactive gases known, right up there with chlorine gas. Any oxygen that formed would have immediately reacted with any metals around to form what chemists call metal oxides (like rust). The composition of the atmosphere back then was probably quite similar to the composition of the atmosphere on the planet Venus. However, unlike Venus, our planet was able to cool to below 212°F, the boiling point of water, and once it did, the gaseous water in the atmosphere and any water locked in the earth's crust liquefied and began to form our oceans. It is estimated that there are about 300 million cubic miles of liquid water on the surface of the earth. We are a "wet" planet. So, anyone who does not believe that water makes an important contribution to our climate is delusional. Various soluble salts from the earth's crust began to dissolve in the oceans. The oceans became "salty."

Let us compare Earth to the planet Venus. As far as we know, the temperature on the planet Venus has never gone below the boiling point of water. Venus has no oceans. Today, the atmosphere on Venus is about 98% carbon dioxide. It would be impossible for the CO_2 level on Earth to go that high with as much liquid water as there is on Earth. There is essentially no water, liquid or otherwise on Venus (less than 0.002%). Venus is a greenhouse gas disaster.

One of the most important properties of carbon dioxide is its ability to dissolve in water. It is one of the most water-soluble gases known. In fact, it is so soluble that it can make water fizz. The carbon dioxide in the atmosphere began to dissolve in the oceans. Some of the dissolved CO_2 remained as soluble carbon dioxide; however, since CO_2 is an acid, a good portion of it reacted with the minerals in the water to form insoluble carbonates. Marble, for example, is calcium carbonate. These insoluble carbonates settled to the bottom of the oceans and eventually became rock. The percentage of CO_2 in the atmosphere began to drop as the gas dissolved. It is important to understand that the solubility of CO_2 in water is not a haphazard, random event, but is controlled by laws of physical chemistry. The amount of CO_2 that will dissolve in water depends on at least two major factors – the pressure of the gas above the liquid, and more important to us, the temperature of the liquid.

These conditions of solubility are formulated in a law of nature, known as Henry's Law, a very simple idea that states that the solubility of a gas in a liquid is proportional to the pressure of the gas above the liquid. As long as the temperature of the liquid remains constant, the relationship between the amount of gas dissolved and the pressure of the gas above the liquid remains linear. That means, for example, if the pressure of the gas above the liquid doubles, the concentration of the gas dissolved doubles. It is important to understand this, because the relationship between atmospheric CO_2 levels and global temperature appears, for the most part, to be linear, suggesting that Henry's Law is involved here.

As a liquid warms and cools, the Henry's Law proportionality changes. Gases are less soluble in warm liquids and more soluble in cold liquids, as anyone who has opened a warm bottle of beer or soda pop knows. Over millions of years, the level of CO_2 in the atmosphere dropped to its present miniscule level of 0.04%. So, as the oceans warm, less CO_2 dissolves and the level of CO_2 in the atmosphere goes up. As the oceans cool, the solubility of CO_2 increases, and the level of CO_2 in the atmosphere goes down. We now can see a major flaw in Vice President Gore's reasoning described in "An Inconvenient Truth." Mr. Gore said, "When there is more carbon dioxide, the temperature gets warmer." What he should have said is, "When the temperature gets warmer, there is more carbon dioxide." Mr. Gore mixed the cause and effect, a typical mistake made by untrained scientists. A very common thing for untrained individuals to do with scientific data is to have the tail wag the dog.

We hear over and over that CO_2 is a greenhouse gas and greenhouse gases regulate the temperature of the planet. First of all, what regulates the temperature of the planet is not a true greenhouse gas effect. The mechanism in an actual greenhouse is quite different from what our atmosphere is doing. So, what in the world is a greenhouse gas?

We do not receive any heat energy directly from the sun. Heat cannot travel through the vacuum of space. We only receive light from the sun. But light energy spans a wide range of wavelengths, from very short, high-energy radiation, such as gamma rays and X-rays, to very long, low energy radiation, such as microwaves and radio waves. "Visible" light wavelengths are about in the middle of this spectrum, a very narrow band of radiation that, due to a quirk of evolution, allows us to see. Just below visible light is ultraviolet radiation at higher energy, and just above visible light is infrared radiation at slightly lower energy. The light spectrum is shown in Figure 3-1. Light of different wavelengths interacts with molecules of matter in different ways. For example, high-energy light like X-rays passes directly through most materials, similar to wind blowing through office windows. Too much of it will really mess the place up. Window glass is transparent to visible light, but absorbs ultraviolet light very strongly. Visible light, the most intense radiation we get from the sun, interacts with matter and, for the most part, is why we see things in color. Infrared radiation, a band of radiation just past the visible

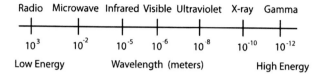

Figure 3-1. Electromagnetic radiation spectrum.

red wavelengths, causes atoms in things to jiggle and generate heat. This is how "heat" travels through the vacuum of space. Understand that molecules absorb light radiation only at very specific wavelengths. Whether molecules can absorb infrared radiation is complicated, but the simple answer is that molecules must be made up of different elements to absorb this form of radiation. There are a few exceptions. More will be said about this in Chapter 6. Microwaves, having slightly longer wavelengths than infrared and a little less energy, cause molecules of matter to spin around, and this also generates heat. Your microwave oven operates this way.

When green plants evolved and began to use the sun's energy in photosynthesis, they began to produce a pollutant waste product known as oxygen, O_2. As we mentioned above, there was no oxygen originally in our atmosphere. The process of photosynthesis ultimately took copious amounts of carbon dioxide from the atmosphere and turned it into our fossil fuels of today, producing O_2 along the way. So, when we burn a fossil fuel, we are simply putting the CO_2 back into the atmosphere, where, apparently, nature intends it to be.

It has been proposed that it took thousands of years before there was a significant amount of free oxygen in the atmosphere. This is because oxygen is so reactive that any oxygen in the atmosphere would have immediately reacted with metals in the earth's crust to form mineral oxides. However, the level of O_2 in the atmosphere eventually built up to where it is today, around 21%.

The earth's original atmosphere did contain ammonia, a gas made up of hydrogen and, more importantly, nitrogen. One theory is that oxygen reacted with the ammonia to form free nitrogen gas and water. Nitrogen is a relatively benign gas that makes up about 78% of the atmosphere. It is slow to react with other substances at ambient temperatures and dilutes the oxygen in the atmosphere to the point where we can tolerate it. Since oxygen, O_2, and nitrogen, N_2, are each molecules containing two of the same element, neither of these gases interact with infrared light and consequently are not greenhouse gases.

The most important greenhouse gas in our atmosphere today is water vapor. It is mainly responsible for keeping the temperature of the globe in check. That should not surprise us considering the fact that about 70% of the surface of the earth is covered with water. Water vapor makes up about 4% of the atmosphere, a hundred times greater than the CO_2 in the atmosphere; however, the level changes from day to day due to weather patterns. Obviously, in dry regions such as deserts, there is essentially no water vapor in the atmosphere. Other greenhouse gases of much lesser importance include carbon dioxide (CO_2), methane (CH_4), nitrogen dioxide (NO_2), and trace amounts of other gases. Some of these gases, like methane, are strong infrared absorbers, and so even at their very low levels, they can contribute a large greenhouse gas effect.

Carbon dioxide could be a very effective greenhouse gas, but, contrary to popular and erroneous belief, it is not very effective in our atmosphere today for a couple of reasons. Carbon dioxide is a strong infrared absorber, but absorbs this radiation over a very narrow band of wavelengths. Carbon dioxide has essentially four ways for the carbon and oxygen atoms in the molecule to jiggle. One of the four ways does not even absorb infrared light, for reasons we need not mention, so only three ways are infrared active. Also, two of the remaining three absorb at a much lower energy than the remaining one. So, unlike water, which also has four ways it can vibrate, all four of them being active absorbers of infrared light, carbon dioxide really has only one strong active absorber of infrared light and a second weaker absorption band. The gas does not trap this energy. After absorbing the infrared light, the molecules then slowly give back this heat energy to the atmosphere, when they bump into other molecules in the atmosphere. This is how the greenhouse gas effect works in our atmosphere, which is quite different from the process in an actual greenhouse.

Now, Earth is a relatively cool planet. So we ask a very important question: how much infrared radiation is available to be absorbed in the first place? Individuals who believe increasing the level of CO_2 in the atmosphere will increase the greenhouse gas effect must assume that the availability of infrared radiation is limitless. But, like the absorption of any type of light radiation, there has to be a source of infrared radiation (heat) at those character-istic wavelengths for a greenhouse gas to be effective. It is similar to color. When there is no visible light present, there is no color. It is difficult to tell what color shirt a person is wearing when it is pitch dark. Another reason to suspect that CO_2 is not an active player in this game is that when you examine the greenhouse gas relationship between atmospheric CO_2 and global temperature going back thousands of years, you find that the relationship is nearly linear. We will explore this idea in a later chapter, but know this to be true. As the level of an absorbing material in a medium increases, its ability to absorb light falls off exponentially. Consequently, for these two reasons alone, the cause of the present up-wiggle in temperature cannot be because of increased CO_2 levels in the atmosphere.

But let's not stop here. We can test these ideas experimentally. We should be able to demonstrate that if CO_2 were driving global temperature, the relationship between the two would behave in a very specific way. In the next chapter, we will devise several simple experiments to see if what we propose as theory is consistent with what we have actually observed in the past 150 years.

4

HOW CAN WE STUDY THE GREENHOUSE GAS EFFECT?

In the previous chapter, we pointed out that most scientists agree the greenhouse gas effect is responsible for keeping the temperature of the globe relatively constant. The question, of course, is: if the temperature of the globe goes wild, that is, if the temperature changes enough so as to cause a major climate change, is that necessarily a consequence of something abnormally wrong with greenhouse gas levels?

Let us explore these ideas by trying some simple thought experiments, although you can actually perform these simple experiments at home. They would be a great science project for your children. For all practical purposes, the greenhouse gas effect is what scientists call a steady-state condition, similar to running water into a bathroom sink, and at the same time, allowing water to run out the drain. If one adjusts the rate of the water coming in so that it is the same as the rate of the water running out, the level of the water in the sink will not change. We say that the water in the sink at this point is in a steady state. It's not a new idea. Many phenomena in nature operate this way.

Let us use this "water in the sink" analogy to describe some different greenhouse gas scenarios. It is well known that not much, if any, infrared light is absorbed by the gases in the atmosphere directly from the sun. The majority of the light from the sun that makes it to the surface of the earth is shorter wavelength visible and ultraviolet light, which is absorbed by the earth's surface. The earth then radiates this energy back to space as infrared light (heat). We will assume that the water entering the sink from the tap represents this heat energy from a warm earth and, indirectly, the sun. The water that is slowly draining out will represent the heat energy being dissipated from the earth back into space. The drain stopper in the sink will represent all greenhouse gases and any other mechanism that retard the dissipation of this heat energy. And, finally, the level of the water in the sink will represent global temperature. A simple schematic of this idea is shown in Figure 4-1.

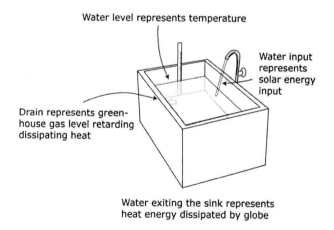

Figure 4-1. Steady-state simulation using a sink to explore the greenhouse gas effect.

With our setup in place, now, let us try several experiments to see if any match the actual behavior of global temperature and greenhouse gas levels over the past 150 years, as illustrated in Figure 2-2. Remember that you can actually try these experiments on your own, if you do not believe the outcomes described here.

Experiment I.

We first adjust the tap and drain stopper so that the water level in the sink does not change. The temperature of the globe, if you will, is constant. At some point, we then close the drain stopper slightly to some new, fixed point and observe the level of the water in the sink. This would be analogous to having the greenhouse gases in our atmosphere increase to some new value and remain there. What will the water in the sink do? As illustrated in Figure 4-2, the water level in the sink will at first remain constant, and then, when the drain is closed slightly to some new, fixed point, begin to rise at a constant rate. Translating this to the greenhouse gas effect, if greenhouse gases

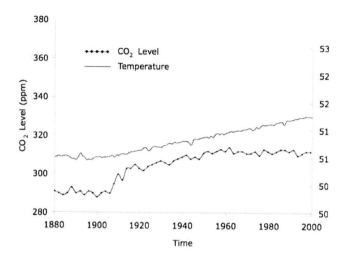

Figure 4-2. Experiment I.

in the atmosphere increased to some new, fixed value with no change in energy from the sun, global temperature would begin to rise and continue to rise at a constant rate. Is this what we observe actually occurring?

Experiment II.

We again adjust the tap and drain stopper so that the water level in the sink does not change. The temperature of the globe is constant. We then begin to close the drain stopper at a slow, but continuous rate. This is analogous to having the greenhouse gas levels begin to increase and continue to increase at a constant rate, which is very similar to the behavior greenhouse gases in our atmosphere today. If you try this with your bathroom sink, you will notice that the level of the water in the sink rises at a faster and faster rate. As illustrated in Figure 4-3, if greenhouse gases were responsible for global warming, global temperature would increase at a faster and faster rate and ultimately pass the increase in greenhouse gas levels. Is this what we observe actually occurring?

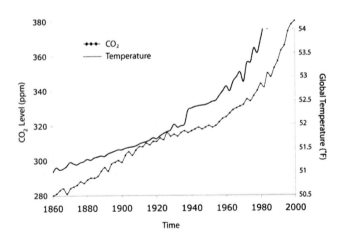

Figure 4-3. Experiment II.

Is it possible to come up with an experiment in which one continuously closes the drain stopper while the water input from the tap remains constant, and the water level in the sink then rises for a short period of time, levels off, and then begins to slowly drop? You can't! It is pretty clear that, without changing the input energy from the sun, the greenhouse gas effect cannot possibly be responsible for the type of global temperature changes that we have been observing for the past 200 years.

Experiment III.

Here is another scenario that might work. Again, we adjust the tap and drain stopper so that the water level in the sink does not change. The temperature of the globe is constant. Now, without changing the drain stopper position, we increase and decrease the tap input periodically. The level of the water in the sink will rise and fall. That is, under the conditions described here, global temperature would increase and decrease periodically, while greenhouse gas levels remain constant. This is now getting close to what we observe, but not exactly. It is true that scientific studies show that over the past 400,000 years global temperature has risen and decreased in a controlled, periodic way. However, what is not happening in this experiment is that the level of greenhouse gases is not changing along with the global temperature, as illustrated in Figure 4-4, but remains constant. That is not what we observe. All the ice-core data shows that as global temperature rises and falls, certain greenhouse gases, one being carbon dioxide, follow along, rising and falling accordingly. The data also show

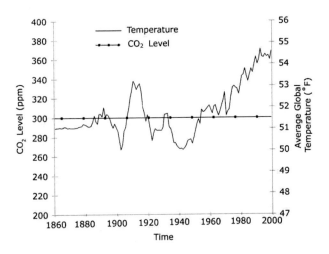

Figure 4-4. Experiment III.

that the CO_2 levels lag behind global temperature, not lead it. Since we cannot find a greenhouse gas scenario that is consistent with these observations, something else must be going on here.

There are several popular erroneous arguments kicking around that support the idea that greenhouse gas emissions are responsible for climate change. Let us put them to rest right now. The first is the premise in Al Gore's "An Inconvenient Truth" that CO_2 levels in the atmosphere cause global temperatures to rise and fall. It's a nice idea, but it does not hold up, when tested experimentally. If CO_2 has been driving global temperature over the last 400,000 years, then what is driving the CO_2?

The relationship that reportedly exists between CO_2 levels in our atmosphere and global temperature is illustrated in Figure 4-5. Please examine the graphs carefully, since they are in part the cause of this whole controversy. It is clear that the two graphs follow each other very closely, too closely to be an accident. Also remember that time

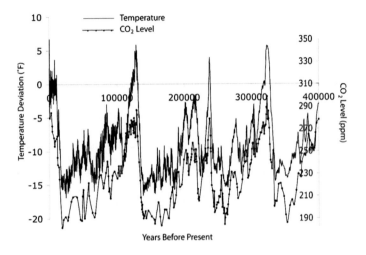

Figure 4-5. Graphs of global temperature and atmospheric CO_2 levels for the past 400,000 years, calculated from Vostok ice-core data. (7) While the graphs appear to exactly overlap, it is clear that the temperature changes before the CO_2 changes.

runs from right to left on the graph. While the two graphs appear to be right on top of each other, it is apparent upon close examination of the graphs that the temperature changes before the CO_2 levels change. So right there, it is impossible for the CO_2 to be controlling global temperature. But let's say that there is some strange type of time warping effect that allows CO_2 to go back in time and control the temperature. Twenty thousand years ago, when the average temperature of the globe was 15°F colder than it is today, and the earth was solidly locked in an ice age, what prompted the CO_2 in the atmosphere to suddenly rise very quickly and take the globe out of the ice age? Perhaps a couple of hundred volcanoes went off in concert, spewing CO_2 into the atmosphere. And, if you believe that, then you would have to believe that, like Old Faithful, all these volcanoes are perfectly timed to go off every 100,000 years. The idea that greenhouse gases control our temperature is not logical. Look at the structure in those two graphs and try to convince yourself that something external, something extraterrestrial, like that big, yellow thing in the sky, is not the culprit here.

Another popular erroneous argument centers around the rate that atmospheric CO_2 is changing. If you ask an "inconvenient truth" supporter why the behavior of CO_2 now is different from the way it behaved over the last 400,000 years, the answer one usually gets is, "Well, we have never seen it go up this fast, and the level of CO_2

is higher now than it has been in millions of years." I am not certain about the "millions of years" part of the argument, but it appears from the graphs in Figure 4-5 that the level of CO_2 is, in fact, higher today than it has been in 400,000 years. So that part of the argument appears to be correct. But is it?

To measure atmospheric CO_2 spanning hundreds of thousands of years, one must find samples of air going back those hundreds of thousands of years. Like the temperature studies, scientists were able to do this by drilling down into Antarctic ice and pulling up ice samples going back thousands and thousands of years. When water freezes, bubbles of air are always trapped in the ice. Consequently, this trapped air can be tested for levels of CO_2. If you look at the original data, one thing jumps out at you. Over a 10,000-year period there are only about a dozen data points, sometimes even less. So between data points, spanning, say, 800 years or so, a lot could be happening to CO_2 levels about which we know nothing. Therefore, it may be true that we have never seen CO_2

levels go up as fast, or to be as high as they are today, but the reason is simple – it is because we have never really looked before. It is only recently that more data is being collected describing the levels of CO_2 in the atmosphere. Moreover, we know that CO_2 is quite soluble in ice as well as liquid water and will diffuse through the ice. Hopefully, those researchers doing all this ice core work took this into account. If they didn't, that could account for the lower levels of CO_2. We now can see why it is so important to carefully define the boundaries of the systems we are studying.

In Chapter 5, we will expand the boundaries of the systems even further. It may be true that CO_2 levels have never been this high in 400,000 years. That is clear from the ice-core data. But we will find that if we go back millions of years, there are only two periods in the globe's history where the CO_2 level in the atmosphere have been *below* 400 ppm. There was a relatively short period 300 million ago where CO_2 levels were below 400 ppm and global temperatures were about what they are today. The second period is the period we are in today. At all other times going back 600 million years, global CO_2 levels have been greater than 400 ppm. So to be concerned because CO_2 levels are nearing 400 ppm is simply not justified based on the history of the globe.

5

IT'S THE OCEANS,
DIM WITS

Let's review! We learned in an earlier chapter that when
the earth formed, the atmosphere was about 90% CO_2
and water vapor. As the earth cooled and eventually the
temperature dropped below the boiling point of water,
the copious amounts of water in the atmosphere rained
on the earth, forming our lakes and rivers, and eventually
our oceans. Any water locked in the earth's crust also
joined the rest. The oceans eventually covered about
70% of the planet's surface. Certain salts making up the
earth's crust are very soluble in water, and eventually
these salts began to dissolve in the rivers, lakes, and

oceans. Rivers and lakes, being much smaller than our oceans, can refresh themselves. The water in these bodies is constantly changing and so it stays "fresh." Our oceans, on the other hand, began to concentrate various salts.

The pressure of CO_2 in the atmosphere was very high, and, as the liquid water cooled, Henry's Law, the relationship between the concentration of a gas dissolved in a liquid and the pressure of the gas above the liquid, went to work, forcing the CO_2 gas in the atmosphere to dissolve in our lakes, rivers, and, more importantly, our oceans. The process was slow, but little by little, the level of CO_2 in the atmosphere dropped. The planet was still very hot, so there was a copious amount of infrared (heat) radiation around, some at CO_2's favorite wavelengths, those wavelengths of light that are absorbed by CO_2 and make it a greenhouse gas. Under these conditions, CO_2 was a very effective greenhouse gas. But as the planet cooled, the pressure of the major greenhouse gas in the atmosphere, water vapor, dropped, causing the surface temperature of the planet to drop even faster.

Carbon dioxide is an acid. As it dissolved, it made our bodies of water acidic. The high concentration of certain salts in our oceans began to react with the dissolved CO_2 to form chemical compounds known as bicarbonates and carbonates. Many carbonates are insoluble in water and began to settle out forming mineral rock such as marble, calcium carbonate. The reaction of CO_2 with the mineral salts removed CO_2 from the water, which, in turn, allowed more carbon dioxide to dissolve. The combination of a number of things – its solubility in water, the reaction of dissolved carbon dioxide with mineral salts, the onset of photosynthesis by green plants – eventually caused the level of CO_2 in the atmosphere to drop into the parts per million range.

If one looks at the data describing the behavior of global temperature and atmospheric CO_2 levels over the past 600 million years, shown in Figure 5-1, one will see something glaringly obvious. It has only been "recently" that CO_2 levels and global temperature show any direct correlation, and there is an important reason for this. For example, during the Cambrian Period 500 million years ago, atmospheric CO_2 averaged around 7000 ppm, about twenty times higher than it is today. Average global

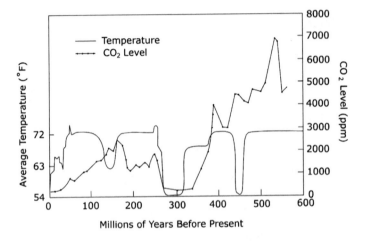

Figure 5-1. Graphs of average global temperature and atmospheric CO_2 levels for the past 600 million years. Temperature graph was constructed using data published C. R. Scotese, University of Texas at Arlington (8); CO_2 graph was constructed using data published by R. A. Berner and Z. Kothavala, Yale University, New Haven (2).

temperature was around 72°F, also about twenty degrees higher than it is today. As the planet headed into what was called the Ordovician Period 450 million years ago, CO_2

levels remained relatively high, 4000 ppm, but average global temperature dropped to about what it is today – no apparent correlation. Then, about 300 million years ago, CO_2 also dropped significantly. Global temperature remained around 54°F, about what it is today, and CO_2 levels dropped to about 300 ppm. This period, known as the late Carboniferous Period, is the only period in the earth's history, except for the present, that atmospheric CO_2 has fallen below 400 ppm. So to expect atmospheric CO_2 to remain down in the 250 ppm range, as those individuals who are screaming that the sky is falling because CO_2 levels have risen to 350 ppm, is not consistent with the "normal" behavior of atmospheric CO_2. The idea that atmospheric CO_2 controls global temperature is simply not supported by any historical data relating the two. If one were to believe that this is true, then, in the Ordovician Period, at a level of 4000 ppm, ten times higher than it is today, global temperature should have been much higher than it is today.

It seems to be clear, then, that once atmospheric CO_2 reaches a particular level, its ability to act as an effective greenhouse gas on this planet disappears. At this point, any greenhouse gas effect is due almost entirely to the water vapor in the atmosphere. We shall explore this idea in more detail in Chapter 6, but the greenhouse gas effect is just not supported by the linear relationship between atmospheric CO_2 and global temperature that we observe experimentally.

Is there a scenario that will describe CO_2 behavior and global temperature behavior over the past 400,000 years, that is a linear relationship that makes sense, and that is consistent with observed scientific data? In fact, we find that there is, and it doesn't involve the greenhouse gas effect at all. Henry's Law, mentioned above, is a fundamental law of nature, a thermodynamic law that describes the solubility of a gas in a liquid. It must operate at all temperatures and pressures. It cannot be repealed by any court of law, or by the wishful thinking of those individuals who find that it is "inconvenient" and doesn't fit into their agenda. The law simply states that the solubility of a gas in a liquid at any temperature is proportional to the pressure of that gas above the liquid. For example, air contains about 21% oxygen, so the pressure of oxygen in the atmosphere (scientists call it the partial pressure) is 21% of the total atmospheric pressure. Oxygen gas is not very soluble in water and so it takes this high pressure to get it to dissolve significantly. The same thing is true for the other major gas in our atmosphere, nitrogen gas, at about 78%. Some gases, like CO_2 and ammonia, actually react with the water when they dissolve, and this increases their solubility, allowing it to be larger than it would be if no chemical interaction took place. These gases are very soluble in water.

If that were all there was to it, then Henry's Law would never be able to explain the relationship between the solubility of CO_2 in water and global temperature. But, as we all have observed when opening a warm bottle of soda, the Henry's Law behavior of a gas in a liquid depends on temperature as well as pressure. Gases are less soluble in warm liquids than they are in cold liquids. The

Henry's Law "constant," the relationship between solubility and pressure, changes with temperature.

So how could global temperature possibly affect the level of CO_2 in the atmosphere? Why when the level of CO_2 in the atmosphere falls below 400 ppm is there an apparent correlation between atmospheric CO_2 and global temperature? Perhaps, this is the wrong question. We may not be defining the boundaries of the system carefully enough. It might make more sense to ask the question, how could ocean temperature affect the level of CO_2 in the atmosphere? In other words, since it appears that atmospheric CO_2 levels follow ocean temperature, then it is logical that the variation of the Henry's Law constant with temperature may account for the comings and goings of atmospheric CO_2 over, at least, the last 400,000 years, where we have the data to observe it.

Let us try the relationship between global temperature and atmospheric CO_2 for the past 150 years or so and see how that compares with observation. The calculations are not difficult; however, one must be careful to account for interactions of the dissolved CO_2 in the ocean with dissolved salts, the concentration of bicarbonate in the ocean, an important salt of CO_2, and the acidity of the ocean. All these are part of the picture and affect the CO_2 after it dissolves. The results are shown in Figure 5-2(a). If we compare the results of this theoretical study with the actual behavior of global temperature and CO_2 over the past 150 years, shown in Figure 5-2(b), we see that the correlation is almost exact. Differences most likely can be explained by the fact that first, global temperature rather than ocean temperature was used, and second, the chemistry of carbonic acid in our oceans is very complicated.

I'm surprised that no one has asked why Henry's Law does not appear to be operating between the temperature and CO_2 levels illustrated in Figure 5-1? It surely looks like temperature and CO_2 levels are all over the place with no correlation at all. It's a good question. It shows that you are thinking. The simple answer is that Henry's Law operates in a linear fashion only when solutions are dilute. This is why it describes the solubility of a

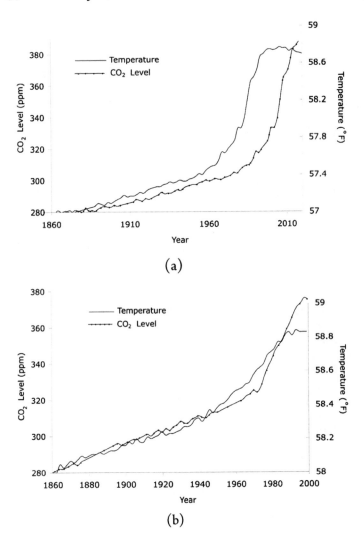

Figure 5-2. (a) Carbon dioxide level versus time as a function of global temperature, calculated using Henry's Law. (b) Actual measured relationship between CO_2 level and global temperature.

gas in a liquid so well. Gases are simply not that soluble in liquids. For example, to get enough oxygen gas to dissolve in our blood so we can live, our blood has a chemical compound called hemoglobin that "grabs" on to the oxygen gas molecules. Without the hemoglobin we would suffocate.

When the concentration of a solution gets high, Henry's Law begins to fall apart. We have said a number of times that CO_2 is unusual as a gas, because it is so soluble in water. It is able to dissolve to a much higher extent, much higher than most gases. While I cannot prove this, my best guess is that once the level of CO_2 above water reaches 1000 ppm, Henry's Law no longer operates in a linear fashion. It seems that as long the level of CO_2 remains down around 400 ppm, Henry's Law is alive and well.

Does this prove that Henry's Law behavior is responsible for the antics of atmospheric CO_2 over the years? No more than one can prove that greenhouse gas emissions are responsible for global warming. That's not the way science works! What we can say, though, is that the Henry's Law explanation is more consistent with observed behavior, and it is simpler and more logical than the greenhouse gas effect, since atmospheric CO_2 levels lag behind global (ocean) temperature, not lead it.

One fact is glaringly true. If the greenhouse gas effect were responsible for global warming, the atmosphere would be heating up. So far, no "hot spots" in the atmosphere have been found. Also, we do know that over the past 200-year period, the oceans have been warming (it's most likely a sun thing), much to the dismay of the polar bears. Scientists have been screaming about melting ice caps for years. The ice-core graphs show that this cycle of warming and cooling about a degree or so occurs over and over again, thousands upon thousands of times. We could easily have been in the cooling part of the cycle during the last 200 years. Then, I suppose, "An Inconvenient Truth" would have been all about the impending doom because there was too little CO_2 in the atmosphere. Actually to be truthful, if you examine the level of CO_2 in the atmosphere over millions of years, shown in Figure 5-1, you would have to admit that the atmosphere presently is starving for CO_2.

A major argument used by global temperature alarmists to justify the so-called "unusual" behavior of atmospheric CO_2 is that we have never seen CO_2 levels go up this fast and this high before. However, if you look at the ice-core data over the past few 100,000 years, you will find that over, say, a 40,000-year period only about 13 samples of the atmosphere were taken. That's the way ice-core data works. So, when you sample CO_2 levels about 3 times in a 10,000 period, there could be a lot going on with CO_2 levels between data points

And even with the little data that we do have historically, if one looks at the CO_2 levels versus time, based on Vostok ice-core data over the past 400,000 years, as shown in Figure 4-5, it appears that the CO_2 levels in the atmosphere were increasing as fast during the period 40,000 years ago when we came out of that ice age, as they are today. The only difference is that global warming advocates decided to continue the upward trend of the CO_2 levels artificially to support their untested theories. There is absolutely no valid scientific reason to do this. This is a perfect example of fixing the boundaries of the system one is studying to coincide with a particular agenda, and is the basic foundation of what is generally referred to as "junk science."

6

WHEN IS A GREENHOUSE GAS NOT A GREENHOUSE GAS?

Those individuals who believe humans are responsible for global climate change like to keep things simple: "Carbon dioxide is a greenhouse gas! Greenhouse gases warm the atmosphere! Beginning in the mid-1800s, with the start of the industrial revolution, humans began to load the atmosphere with carbon dioxide! Global temperature began to rise significantly at that point! Humans are causing the temperature of the globe to go up!" You can almost hear the drum accent after each of those statements. Wouldn't it be nice if nature operated in such a simple and theoretical way? If this were the case, we probably would be able to predict the weather accurately a week in advance. In this chapter, we must get a little more technical. We need to address the fact that over the past 400,000 years the relationship between global temperature and CO_2 level appears to be linear or nearly linear. We will see in this chapter that the absorption of light by matter, infrared or otherwise, is not a linear relationship with concentration.

To begin, I will pose a series of simple questions to raise the wattage of our brains: Is carbon dioxide a greenhouse gas? Is carbon dioxide always a greenhouse gas? What makes carbon dioxide a greenhouse gas? In the absence of heat (infrared) radiation, would carbon dioxide be a greenhouse gas? Is it possible for carbon dioxide to be an effective greenhouse gas on one planet and not on another? When is a greenhouse gas not a greenhouse gas? We see that it's not so simple to say, "Carbon dioxide is a greenhouse gas and that's a fact!"

Asking when a greenhouse gas is not a greenhouse gas is the same thing as asking when a color is not a color?

The ability to have color and the ability to be a greenhouse gas are properties that physical scientists refer to as contextual properties. These are properties that do not entirely belong to the material in question. Let us examine this property that we call color first, since it is easier to understand than the ability to be a greenhouse gas. When you place a "green" sweater in a drawer and close the drawer is the sweater still "green"? I am sure that the average person not trained in physics or chemistry would say, "Don't be ridiculous! Of course it is." But is it? This property we call color depends on at least three variables. First, the color of an object depends on some pigment or dye that gives the object a particular color. But that's not enough. The property also depends on a source of illumination (the sun, a light bulb, a flame, etc.), and a detector to detect the radiation reflected from the object (human eye, a camera, etc.). Take any one of these things away and color does not exist. So, in the absence of visible light, the color "green" does not exist. When you place that "green" sweater in a drawer and close the drawer, the sweater actually has no color. If one fiddles with any of the variables described above, the source of illumination for example, one could actually change the color. How many of us have purchased an item of clothing at a store thinking it was a particular color, only to find when we were out in the sunlight, it was a different color. Color is a contextual property.

The ability of a gas to be a so-called greenhouse gas also is a contextual property, since it involves the absorption of light. It too depends on variables that do not belong to the gas. For example, it is not necessarily true that if the concentration of CO_2 in the atmosphere doubled, the ability of CO_2 to absorb radiation would also double. As mentioned in a previous chapter, the energy we get from the sun comes to us only in the form of light. For most people, the term "light" usually is understood to be visible light, the radiation we detect with our eyes. But "light" (scientists refer to it generally as electromagnetic radiation) is a wave phenomenon that covers a wide variety of wavelengths (see Figure 3-1).

A wave is a disturbance that travels through space. It is usually associated with something (a rope, water, a guitar string, etc.) wiggling up and down, but it also could be a single impulse, such as a shock wave from an explosion. The wiggly type of waves are characterized by at least two parameters, *frequency*, which is the number of times the medium carrying the wave wiggles up and down per second, and *wavelength*, the distance between equivalent positions along the wave. These two parameters are related to each other. When the wavelength goes up, the frequency goes down and vice versa. Waves move through space at a velocity that is equal to the wavelength multiplied by the frequency. For electromagnetic waves, this velocity is the speed of light, 186,000 miles per second.

The energy of the light radiation increases when the frequency of the light wave increases, but decreases when the wavelength increases. For example, very high-energy light, such as gamma radiation or X-radiation, has a high frequency and a very short wavelength. This high-energy radiation is very dangerous to biological cells. Very low energy light, such as microwaves or radio waves, have a low frequency or long wavelength. As far as we know, this low-energy light is benign. Visible light, the radiation we detect with our eyes, is about in the middle of these extremes, and is the primary radiation earth receives from our sun. All these frequencies or wavelengths make up what normally is called the *light spectrum.*

Molecules, the fundamental building blocks of matter, absorb this electromagnetic radiation. The type of radiation absorbed by molecules depends on a number of complicated factors that are not important to mention here, and interacts with the molecules in different ways. Microwaves, for example, cause molecules to spin around faster. Infrared radiation, the radiation having wavelengths slightly longer than visible light, causes the atoms in molecules to jiggle or vibrate faster. We perceive this spinning or vibrational energy as heat energy. A molecule must be able to absorb this heat energy to be a greenhouse gas.

A large portion of the radiation produced by the sun is filtered out by our atmosphere and never reaches the earth's surface. Another large portion of light is reflected off the surface of the planet without being absorbed. If you were on Mars, Earth would appear as a shining star. Without our atmosphere, and the important molecules in it, such as CO_2 gas, plant and animal life on the planet would literally be toast. Even now, with an atmosphere, we are being strongly advised to stay out of the sun. The majority of light radiation from the sun that makes it to the surface of the earth falls in the visible region of the spectrum. Scientists agree that this is most likely the reason we see in this region of the spectrum. As we described above, this visible light is, in fact, the radiation that gives objects color. The gases in our atmosphere absorb very little visible light. Our atmosphere essentially is colorless. Once the Earth absorbs this radiation, it then slowly releases this energy back to the atmosphere, a good part of it in the form of infrared, or heat, radiation.

Certain gases in our atmosphere will absorb this infrared radiation. Why some gases absorb this radiation and others do not is complicated, but the bottom line is that molecules that contain more than two atoms, such as carbon dioxide (CO_2 – 3 atoms), water vapor (H_2O – 3 atoms), methane (CH_4 – 5 atoms), etc. do absorb this radiation at very specific wavelengths. Molecules with only two atoms will absorb infrared radiation only if the two atoms are different elements, for example, carbon monoxide (CO), and nitrogen monoxide (NO). The two most abundant molecules that make up our atmosphere, oxygen gas (O_2) and nitrogen gas (N_2) are not infrared active, since both atoms in each of these molecules are identical. So, like color, the ability to be a greenhouse gas requires a source of radiation. The ability of molecules such as CO_2 and water vapor to absorb this infrared or heat energy and transfer it to other molecules in the atmosphere is what we refer to as the "greenhouse gas effect." Without these greenhouse gases the planet would be uncomfortably cold.

Carbon dioxide absorbs infrared radiation at three specific or favorite wavelengths. One of these three wavelengths absorbs more strongly than the other two, and the two weaker modes actually absorb at the same wavelength, so the gas essentially has only two absorption bands. This ability to absorb heat radiation is similar to a colored object having some sort of pigment to give it the color.

The amount of radiation absorbed by gaseous molecules depends on, among other things, the concentration of that gas in the system, but the relationship is not linear. Relatively speaking, the concentrations of greenhouse gases in our atmosphere are very small, but, apparently, just right enough to keep the planet hospitable for short periods of time. As we said, the most abundant gases in the atmosphere are nitrogen and oxygen making up about 99% of the atmosphere. All other gases are, for the most part, present in trace amounts. The level of water vapor, for example, varies, but averages around 4%. This represents 400 people in a group of 10,000 people. The level of carbon dioxide is one hundred times smaller at around 0.04%, which is 4 people in 10,000 people. Methane and other greenhouse gas concentrations are even smaller. One recent proposal of some individuals is the idea that if we could lower the level of CO_2 in the atmosphere from 380 ppm (0.038 %) to 350 ppm (0.035%), that will hold off global climate change. This is a change of 0.003 %, which is equivalent to less than 1 person in 10,000. My guess is that there are regions on the globe where CO_2 levels change by this amount on a daily basis, around volcanoes, for example. With the "right" level of greenhouse gases to keep the planet hospitable, the planet has still periodically been much warmer than it presently is today, and much cooler, very much cooler, than it is today. In fact, the earth has gone into an ice age with carbon dioxide levels in the atmosphere much higher than they are today. How is that possible? According to those who believe that carbon emissions are causing the planet to warm, it is not.

To understand this, we need to go back and examine color again. A particular color exists only if the illumination source has available certain wavelengths of visible light. Take those wavelengths away and that particular color is not possible. Let us assume that the wavelengths of light that allow an object to appear red are absent from the source of illumination. The object will not appear to be red. Moreover, you can attempt to "color" it with as much red pigment as you want and it still will not appear to be red. The same thing is true for greenhouse gases to be greenhouse gases. For carbon dioxide to be an effective greenhouse gas, infrared radiation of very specific wavelengths must be present. If they are not, the gas is not a greenhouse gas, or is a poor greenhouse gas.

The sun produces only so much radiation that reaches the earth. The planet is a relatively cool planet as planets go. That is, it produces infrared radiation, but the intensity of the radiation at those favorite wavelengths of CO_2 is relatively low. As the level of carbon dioxide in the atmosphere increases, it absorbs more and more of this radiation. But that is not all there is to it. As the level of

CO_2 in the atmosphere builds linearly, the amount of radiation absorbed by the CO_2 does not increase linearly.

Think of it this way. Place a piece of paper in front of a light bulb. The paper reduces the intensity of the light. Add another piece of paper. The intensity of the light drops, but to a smaller extent. Add another sheet of paper. The intensity of the light further decreases, but to an even smaller extent. At some point adding another sheet of paper will have no additional effect. Essentially all the available light is absorbed. However, if you increase the intensity of the light bulb, say going from 60 watts to 100 watts, it will take more paper before the absorption is complete.

One will find that the intensity of the light through the paper drops logarithmically as each piece of paper is added. This relationship of light absorption has been known for a long time and is called Beer's Law. What amazes me is that many climate scientists do not think it applies to the absorption of infrared light. Carbon dioxide is the piece of paper. We must be careful, though, not to push this analogy too far. Beer's Law relates the drop in intensity of light through a medium to two parameters, the concentration of the absorbing substance in the medium and the path length of the light through the medium. Assuming all the pieces of paper are identical, the above analogy focuses on the path length of the light rather than the concentration of the absorbing substance.

The greenhouse gas effect is a concentration effect, not a path length effect. The intensity of light drops because the number of CO_2 molecules is increasing in a constant path length. A better analogy might be to consider passing light through a colored solution in which the concentration of the dye in the solution is increasing. In any case path length or concentration, in order to see a linear increase in global temperature, CO_2 levels in the atmosphere would have to increase exponentially. The ice-core data going back thousands of years simply does not support this, and, I believe, this is the nail in the coffin. Climate change cannot be a consequence of the greenhouse gas effect.

Some scientists refer to the ideas mentioned above as "radiation saturation," an unfortunate and misleading term, since it implies that the gas is saturated with radiation. In fact, it is just the other way around. I like to think of this as "radiation depletion." At some point there is not enough radiation to saturate the gas. Once the available radiation has been absorbed, adding more carbon dioxide to the system has no effect. The question, of course, is: at what level of CO_2 does this "depletion" effect takes place, because once it does, carbon dioxide ceases to be an effective greenhouse gas. The exact level of CO_2 is not known, and since the relationship is logarithmic, the gas never truly "saturates." Just like the intensity of the light bulb described above, the intensity of the incident infrared radiation will affect the level at which the gas is for all practical purposes "saturated." Calculations by various scientists, including myself, seem to point to an effective saturation value on this planet of around 300 to 400 ppm. If this is true, then at the present level of CO_2 in the atmosphere, the gas has pretty much absorbed all the available infrared radiation at its favorite wavelengths. Would this be the case on another planet, such as Venus? Venus is closer to the sun and is a much hotter planet. Consequently, it should take a much higher CO_2 level to saturate the gas.

Some scientists will argue that as the gas reaches its saturation point, the band of wavelengths it can absorb broadens, and so the gas also will absorb additional radiation at wavelengths close to its favorites. While this certainly may be the case, calculations show that the effect is minimal. The radiation at wavelengths close to favorites also will be depleted. Others argue that some gases such as CO_2 encourage other gases to absorb more radiation. This also is a stretch. It's apparent that the sun does not produce more radiation, simply because the CO_2 level in the atmosphere increases. How do we know this? If this were the case, the earth most probably would never have cooled to below the boiling point of water. We would really be the sister planet to Venus.

If one believes that the ice-core data here on Earth is correct, and there is no reason to assume that it is not, then, looking back over millions of years, it is evident that the earth has survived "runaway" greenhouse gas effects before. But even more important is the fact that the earth has gone into ice ages with CO_2 levels much higher than they are today. It seems that on Earth once the greenhouse gas CO_2 reaches a level of around 300 ppm in our atmosphere, increasing its concentration has little additional effect on the temperature of the globe. This behavior is illustrated in Figure 6-1. The data plotted on the graph is a consequence of the relationship between light passing through an absorbing substance and the concentration of that absorbing substance. This relationship is described by a law in optics known as Beer's Law (sometimes called the Beer – Lambert Law).

We can see from the graph that the majority of the radiation is absorbed at the low levels of the gas and as the concentration of the absorbing substance increases, its effectiveness decreases. Actually, this loss in the ability to

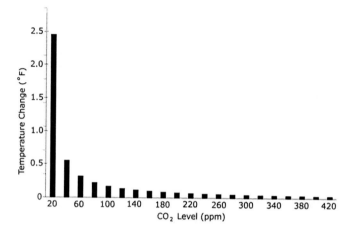

Figure 6-1. Effectiveness of carbon dioxide as a greenhouse gas as a function of atmospheric carbon dioxide level, as predicted by Beer's Law. (calculations by J. R. Barrante, *unpublished data*).

absorb radiation at increased concentrations has more to do with the incident radiation than it has to do with the CO_2. As the radiation is absorbed by the CO_2 at specific wavelengths, there is a depletion effect. Think of it as a sponge absorbing water. Once the water is absorbed, adding more sponges will have no additional effect. Consequently, increasing the CO_2 from its pre-industrial level of 280 ppm to the present level of 380 ppm has essentially no effect on global temperature, again supporting the contention that the present global warming is not due to carbon emissions. More will be said about this depletion effect in Chapter 7. Along with the fact that the CO_2 follows the global temperature, not leads it, this depletion effect is one of the most important pieces of evidence supporting the thesis that greenhouse gases cannot be responsible for climate change.

7

FILLING IN
THE BLANKS

In a previous chapter we pointed out that one of the ways scientists measured atmospheric CO_2 going back thousands of years was by testing air trapped in core samples of ice from places like Vostok Station, Antarctica. The problem with this method is that the samples of air are removed from wide bands of ice, and so there is a large uncertainty as to the exact date of the ice. Moreover, to remove the air from the ice, one must melt the ice, and this could cause the water that forms to de-gas, since we know that CO_2 is soluble to a certain extent in ice. Consequently, over an 18,000 year period there could be as few as 14 samples. That represents one sample in every thousand or so years. Figure 7-1 shows CO_2 levels determined from an actual portion of the Vostok ice-core data for an 18,000-year period. One will note that with only 14 data points, a lot could be happening with the CO_2 levels between the data points.

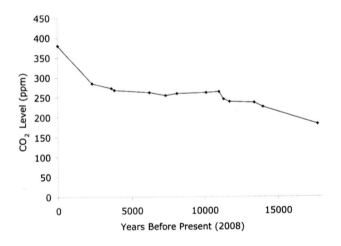

Figure 7-1. Level of atmospheric CO_2 as a function of time for past 18,000 years, calculated from Vostok ice-core data. (7) Note that there are only 6 data points between 11,000 years ago and 18,000 years ago, when the planet came out of the last ice age.

We showed in Chapter 2 that Henry's Law, the law that relates the solubility of a gas in a liquid to the gas pressure (and consequently the concentration) above the liquid, describes the relationship of atmospheric CO_2 to its solubility in the oceans quite well. The discrepancies between actual and calculated CO_2 levels are because of the fact that the chemistry of carbon dioxide, once it dissolves, is quite complicated and difficult to model exactly with simple laws. Furthermore, and this is important, we only have global air temperatures and not exact ocean temperatures to work with. It is clear that the rapid swings in the temperature of the atmosphere would be buffered to some extent in the oceans. Water just doesn't warm and cool as rapidly as air. Also, once water freezes, the layer of ice that forms insulates the water below it. Consequently, air temperature will go much lower than water temperature. So, we will see artificially lower values for CO_2 levels than are actually measured, especially at low temperatures. It would be nice to have accurate ocean temperatures going back thousands of years, but we don't. We can only guess.

Nevertheless, we can use Henry's Law to help us find

out what is happening to the CO_2 levels between those

data points. Before we do this, however, in order to be

scientifically honest, let's see how we do with information

we know to be correct. Figure 7-2 shows a comparison of

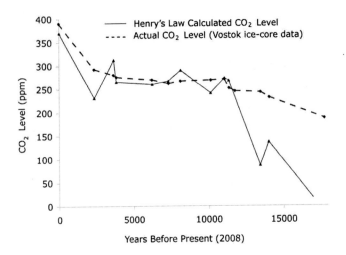

Figure 7-2. Comparison of atmospheric CO_2 levels, calculated from Vostok ice-core data, to those levels, calculated using Henry's Law. Note the large discrepancy between calculated and observed CO_2 levels from 12,000 to 18,000 years before present. This is because atmospheric temperatures rather than ocean temperatures were used.

atmospheric CO_2 levels as calculated with Henry's Law (with the same number of data points as the measured levels) to the actual data shown in Figure 7-1. Note that the graphs are quite similar at high temperature, but deviate significantly at low temperatures between 12,000 years ago and 18,000 years ago, as we expected would happen. This is because air temperature rather than ocean temperature was used to do the calculations.

However, even with these discrepancies we can see that Henry's Law shows the CO_2 levels rapidly rising 18,000 years ago, when we came out of the last ice age and oceans began to warm. If we assumed that carbon dioxide was controlling the climate, then what triggered the carbon dioxide to "suddenly" begin to increase? No one was burning fossil fuels 18,000 years ago, and there were not many SUV's around. Consequently, instead of the tail wagging the dog, as it does in the documentary "An Inconvenient Truth," let us try having the dog wag its tail. It is much more likely that the oceans warmed first and then the CO_2 levels began to go up. This is absolutely consistent with Henry's Law.

We now can fill in the blanks. Let us now use Henry's Law to calculate atmospheric CO_2 levels between those data points, in order to see what possibly could be happening between the measured points. Remember, we are still using air temperature rather than ocean temperature to do the calculations. The results are shown in Figure 7-3.

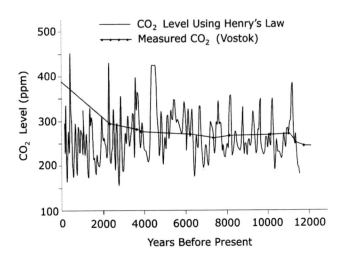

Figure 7-3. Atmospheric CO_2 level calculated as a function of global temperature for the past 12,000 years using Henry's Law, with a comparison to the measured values from ice-core data. (7)

We can see that between the measured data points, the CO_2 level changes significantly. Moreover, temporary spikes in the level (perhaps lasting 100 years or so) over 400 ppm are quite common. Also, Henry's Law shows that the average CO_2 level stayed relatively constant for the past 2000 years and then suddenly began one of its upward spikes in the last 100 years or so, as the ocean temperature began to increase. This suggests that statements to the effect that the CO_2 has never been as high in the last 1000 years as it is today are not correct.

But it is time to put the nail in the coffin and bury the idea that CO_2 is responsible for climate change. There is a procedure used by chemists to prove whether a mechanism, the detailed series of steps generally occurring at a molecular level that describes how the process takes place, is valid. It is important to understand that the procedure cannot prove that the mechanism is correct. But it definitely can prove that the mechanism is not correct. Sounds confusing? Maybe this will help. A chemist proposes that a particular chemical process takes place in a very specific way, the mechanism for the process. He or she then uses this proposed mechanism to develop a mathematical equation that will tell the chemist how fast the process will take place. The chemist then goes into the laboratory and measures the rate at which the process takes place. If the measured rate agrees with the rate predicted by the proposed mechanism, then the chemist knows that the mechanism could be correct. It doesn't prove that the proposed mechanism is correct, because another mechanism could produce the same rate equation. But, if the measured rate does not agree with the rate predicted by the proposed mechanism, then the chemist knows that the proposed mechanism is definitely wrong.

What does this all have to do with global warming, you ask? Let us use this approach to examine whether carbon dioxide could possibly be responsible for global warming. According to "An Inconvenient Truth," the mechanism for global warming is the greenhouse gas effect. The greenhouse gas effect requires the absorption of light and the theories relating the absorption of light to concentration have been around for a couple of hundred years.

Look at the graphs in Figure 4-5. Notice that the CO_2 levels in the atmosphere follow the temperature changes very closely. The relationship between CO_2 level and temperature looks almost linear. If CO_2 were actually driving the temperature by the greenhouse gas effect, what should we expect the behavior of CO_2 to look like in Figure 7-2?

We will begin with the fact, cited in Mr. Gore's documentary, that the increase in atmospheric CO_2 levels from 280 ppm to 380 ppm caused global temperature to rise about $0.8\,^{\circ}C$ $(1.4\,^{\circ}F)$. This is not consistent with the graphical information shown in Figure 6-1, but we will give him that for the purpose of this discussion. Also, we will assume, as they must have done in the documentary, that the amount of infrared radiation is limitless. That is, there is no radiation depletion effect, which we described in a previous chapter.

It is well known that the relationship between CO_2 level and temperature increase is a simple Beer's Law type relationship. Beer's Law was originally developed to describe the absorption of visible light as it passes through a solution and then related the drop in the intensity of the light to the properties of the solution, such as its concentration. We know now that the absorption of all forms of light, including infrared light, obeys Beer's law. Now the greenhouse gas effect may not follow Beer's Law exactly, but it should be close, since it involves the absorption of infrared light. What is so special about Beer's Law, you ask? It is that the relationship between temperature change and CO_2 level is not linear it is logarithmic. Any of the thousands of scientists making up the "consensus" should know this.

Those not trained in mathematics will have to take my word that the outcomes I propose here are correct. Because Beer's Law is a logarithmic relationship, the law requires, using data presented in "An Inconvenient Truth," that in order to get a temperature change of   approximately 3°F, the concentration of CO_2 in the atmosphere would have to exactly double! Twenty thousand years ago, when we came out of the last ice age, global temperature increased by about 15°F. Consequently, if CO_2 were driving that change, its concentration would have had to double 5 times. Ice-core measurements put the CO_2 level back then at about 150 ppm. For every 3°F change, the level of CO_2 would have had to go from 150 ppm to 300 ppm to 600 ppm to 1200 ppm to 2400 ppm to 4800 ppm. The results are shown in Figure 7-4. If any other greenhouse gas, say water vapor, were responsible for that 15°F rise in temperature, the level of water vapor in the atmosphere would have had to do a similar thing.

But wait! The whole point of this book is that the rise in global temperature of 1.4°F was not caused by an increase in atmospheric CO_2 from 280 ppm to 380 ppm. The Beer's Law depletion effect illustrated in Figure 6-1 shows that the increase in CO_2 level from 280 ppm to 380 ppm would increase global temperature 0.14°F, not 1.4°F. This is because at these levels CO_2 has pretty

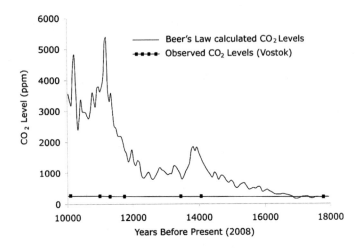

Figure 7-4. Comparison of expected required CO_2 levels calculated using Beer's Law to observed ice-core measured CO_2 levels (1). The graph shown uses "The Inconvenient Truth" data that doubling the CO_2 level increases the temperature by $3°F$. The graph utilizing the data from Figure 6-1 has not been included, since it is not needed to make the point.

much absorbed all the radiation available at its wavelengths. Consequently, doubling the CO_2 level in the atmosphere would raise the temperature of the atmosphere about $0.3°F$. To cause a $15°F$ increase in temperature would require the CO_2 level to double 50 times, not 5 times, giving us an astronomical level of CO_2 in our atmosphere.

So, look at the behavior of CO_2 in Figure 7-1. The level did not even double one time. In fact, there is no evidence what so ever that the level of any greenhouse gas in the earth's atmosphere increased enough to raise the temperature 15°F. Like the chemist's problem described earlier, this does not prove what caused the 15°F rise in temperature, but it clearly proves the idea that green-house gases could be responsible for climate change is definitely wrong. Could any of the greenhouse gases have been responsible for the warming trend 18,000 years ago to some small extent? Absolutely! Certainly water vapor is an important greenhouse gas and most likely contribut-ed its share to the warming trend. But the greenhouse gas effect is simply not a major player in climate change. In fact, the scientific data supports the inverse.

What about the future? That is a good question! Many individuals believe that the most important use of science is to predict the future. While it is important, it also is one of the most dangerous uses of science, since science does much better explaining the past than predicting the future. So, we must say right from the start that any predictions of future climate changes are, at best, only guesses. To believe that a computer model can actually accurately predict the future is one of the most common mistakes individuals have made concerning climate change. Moreover, to get excited over, be concerned about, and spend trillions of dollars trying to alter these predictions is ludicrous. This is evidenced by the fact that most of the predictions made in the documentary "An Inconvenient Truth" did not occur, or, in fact, were wrong. The predictions made about global warming in the original Kyoto Protocol also did not occur, and many also were wrong. Computer models predicted that if the present warming trend was caused by the greenhouse gas effect, there would be a hot spot about 6 miles up in the atmosphere over the tropics. Scientists have been looking for this hot spot for the past dozen years or so. So far, no hot spot has been found. Perhaps the only prediction that was correct was the prediction that atmospheric CO_2 levels would continue to rise, a fairly safe prediction, since atmospheric CO_2 has been rising and falling for millions of years. It is like predicting the weather. They had a 50% chance of getting it right.

So, are we heading for a climate disaster, as many people would have us believe? Based on previous climate changes going back thousands of years, I would have to say that it is extremely likely that we are, since it appears that is what the climate likes to do. Is it possible that we could stop the climate from changing? I wouldn't bet any money on that, no less break the budget of most countries in the world trying to stop it!

We can safely say that in the next few hundred years, unless some catastrophic event, like being hit by an asteroid, occurs, nothing drastic is going to happen to the climate, irrespective of what humans or polar bears do on the planet. Large swings in global temperature take thousands of years to take place. So, global temperatures should wiggle up and down a degree or so as they have been doing for millions of years. These wiggles occur over a 200-year cycle, so it is highly probable that every time we enter one of these up-wiggles, some politician is going to scream that the planet is warming and we have to fix it.

In the past 400,000 years the planet has experienced four ice ages (see Figure 2-1). Moreover, the "warm" periods, similar to what we are experiencing now, occurred only five times in the last 400,000 years and lasted for relatively very short periods of time. In fact, a warm period that occurred approximately 240,000 years ago lasted only about 5000 years, a very short period of time on a global time scale. Out of the last 400,000 years, the globe has spent about 300,000 years at very un-hospitable temperatures. The longer warm periods last about 10,000 years. If you examine the period we are presently in, you will see that our time is just about up. Global warming aside, we have been in this period for about 10,000 years, and, based on her previous behavior, Mother Nature is just itching to get back to ice age.

So, let us do some "junk science" and predict the future. We know that the climate changes very slowly. The last climate change took place over a period of about 8000 years. We also know that the present warming trend and spike in CO_2 levels is about 150 years into its 200-year cycle. It is highly probable that within the next 50 years we will begin to see a turn around in global temperature as we start the downward part of the cycle. Since the CO_2 levels lag behind ocean temperature several hundred years, CO_2 levels will continue to rise (whether we do anything or not) for the next few hundred years and then the atmospheric CO_2 levels will begin to drop, as the oceans begin to cool. What happens next is pure speculation.

Based on the behavior of the climate for past 400,000 years, the events described in Figure 7-5 are likely to occur within the next 1000 years. It is evident from the temperature profile shown in the graph for the last 10,000 years that nothing remarkable stands out. In fact, the graph looks pretty flat. There is no evidence that we are presently experiencing some type of unusual global warming event.

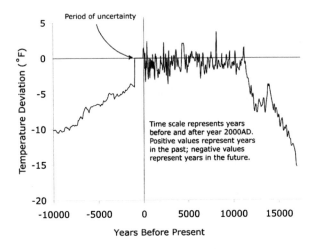

Figure 7-5. Predicted global temperature for the next 10,000 years, based on trends in global temperature for the past 400,000 years. Lack of detail in the future portion of the graph is because we have no detailed information on future temperatures. Since we do not know when this cooling trend will begin, there is a period of uncertainty projected to last for about 1000 years.

Since we do not know exactly when the future-cooling trend will begin, there is an uncertain period of about a thousand years. However, based on the previous behavior of the climate, it is unlikely that we will stay at our present warm period for another thousand years.

These warm periods do not last for more than about 10,000 years, and, as we said previously, our time is up. So, within the next thousand years there is going to be a noticeable drop in temperature. People living away from the equator may or may not notice it first with shorter summers and longer winters. It will happen very slowly. During this period of time, average global temperature should drop a couple of degrees F. No humans will really be aware that the climate is changing. Our lifetime is just too short. Each generation may notice some effects, but ten generations is less than a thousand years, and we have seen that not much happens to the climate in a thousand years.

Over the next six or seven thousand years the temperature will drop another six degrees F. All calculations clearly indicate that by -10,000 on the graph in Figure 7-5 (that would be about the year 12,000), the planet will be in our next ice age. Much of North America will be under glacier ice. The Bering Sea will have frozen over re-establishing a land bridge between Alaska and Siberia.

Polar bears will be celebrating, maybe! It's really going to be cold!

Can we be confident that the future climate will behave this way? We can only guess that it will, based on the fact that it has followed this scenario four times in the last 400,000 years. At present, there is absolutely no reason to assume that it will not. Will there be any difference between our next ice age and the previous ice age 15,000 years ago? One of the most obvious differences will be the level of the human population and its technology. During the last ice age, relatively speaking, there were not many humans around, and those that were had a very limited technology. Our wisest ancestors did the right thing. They adapted to the climate change and moved toward the equator. Those that didn't perished. Hopefully, future humans will be smart enough to survive. Sadly, unless technology can solve food problems, many humans will starve.

Based on previous behavior of the climate, it is safe to say that on the planet Earth global warming will never be a concern, unless something catastrophic happens to our sun. It is true that after the next ice age, the planet will again warm to temperatures close to what we are experiencing today. But, as I hopefully have shown in these chapters, this will not be due to any greenhouse gas effect.

SOURCE MATERIAL
and
SUGGESTED READING

The information listed here has been compiled as source material and suggested reading for those interested in learning more about the history of climate on Earth. Because I suspected it would be easier for individuals to go online rather than visit a university library, several articles have been taken from the Web, which sometimes is looked down upon because it is not peer-reviewed. Believe me, these articles have been peer-reviewed. In any case, the readings below are excellent and are there for your perusal.

(1) Archer, D., *Global Warming: Understanding the Forecast*, Blackwell Publishing, Malden, MA, 2007.

(2) Berner, R. A. and Kothavala, "Geocarb III: A Revised Model of Atmospheric CO_2 Over Phanerozoic Time," *Amer. Journal of Science*, **301**, 182-204 (2001).

(3) Gore, A., *An Inconvenient Truth, The Planetary Emergency of Global Warming and What We Can Do About It*, Rodale Press, Emmaus, PA, 2006.

(4) Hieb, M., "Climate and the Carboniferous Period," *http://www.geocraft.com/WVFossils/Carboniferous_ Climate.html*.

(5) Hoffman, D. L., "Warming Caused by Soot, Not CO_2,"*http://scienceandpublicpolicy.org/images/stories /papers/commentaries/soot.pdf*.

(6) Nelson, T. J., "Cold Facts on Global Warming," *http://brneurosci.org/co2.html*.

(7) Petit, J. R., Jouzel, J., Raynaud, D., Barkov, N. I., Barnola, J. M., Basile, I., Bender, M., Chappellaz, J., Davis, J., Delayque, G., Delmotte, M., Kotlyakov, V. M., Legrand, M., Lipenkov, V., Lorius, C., Pépin, L., Ritz, C., Saltzman, E., Stievenard, M., "Climate and Atmospheric History of the Past 420,000 Years from the Vostok Ice Core," *Nature* **399**, 429-436 (1999).

(8) Scotese, C.R., *http://www.scotese.com/climate.ht*

LIST OF FIGURES

Figure
                                          Page

2-1. Global temperature, 400,000 years   17

2-2. Global temperature, 150 years   23

2-3. Global temperature, 300 years   24

2-4. Global temperature, 1000 years   25

2-5. Global temperature, 5000 years   26

2-6. Global temperature, 20,000 years   28

3-1. Electromagnetic radiation spectrum   37

4-1. Simulation of greenhouse gas effect using sink   45

4-2. Greenhouse gas simulation I   47

4-3. Greenhouse gas simulation II   48

4-4. Greenhouse gas simulation III   50

4-5. Global temperature and CO_2, 400,000 years   52

5-1. Global temperature and CO_2, 600 million years   60

5-2. CO_2 level with Henry's Law, 150 years   66

6-1. Effectiveness of CO_2 as a greenhouse gas   87

7-1. CO_2 as a function of time, 18,000 years   90

7-2. CO_2 calculated with Henry's Law, 18,000 years   92

7-3. CO_2 using Henry's Law, filling in the blanks   94

114   JAMES R. BARRANTE

7-4. Expected CO_2 level using Beer's Law   100

7-5. Predicted global temperature, next 10,000 yrs   106

GLOSSARY

accuracy – The relative error between a measured value and the value that has withstood the test of time. Sometimes confused with the term *precision*.

atom – The smallest, chemically indivisible particle of matter, composed of protons, neutrons, and electrons. It is the number of protons in an atom that identifies the atom as a particular element.

Beer's Law – A law of optics, originally formulated for visible light by August Beer and others in the mid-1800s, that describes the absorption of light passing through a solution (air is a solution). The law requires that the intensity of light passing through the solution must drop off exponentially (logarithmically) as the concentration of the absorbing substance increases. We know now that the laws applies to all forms of light, not only visible light.

bicarbonate – A salt of CO_2 formed by the reaction of carbonic acid with bases such as sodium hydroxide (lye).

Sodium bicarbonate ($NaHCO_3$) is a mild base and therefore is used as an antacid.

boundary of a system – In science, a barrier which could be real or hypothetical that separates a system under experimental investigation from its surroundings. If defined correctly, the investigator can be confident that no outside and perhaps unknown influences on the experiments performed will be present.

carbon – An element containing 6 protons in its atoms. As far as we know, it is a necessary element for life.

carbon dioxide – A non-polluting chemical compound composed of 1 atom of carbon and 2 atoms of oxygen, CO_2. It is a colorless, odorless gas at room temperature formed generally as a combustion product of hydrocarbons. In water it forms an acid H_2CO_3, known as carbonic acid. In solid form it is known as "dry ice." In animals it controls the acidity of the blood.

carbon monoxide – A polluting, very poisonous gaseous compound composed of 1 atom of carbon and 1 atom of oxygen. It is formed in the incomplete combustion of hydrocarbons. In animals it blocks the ability of the blood to take up oxygen and causes asphyxiation.

carbonates – Salts of CO_2 formed from the reaction of carbonic acid with mineral bases such as calcium oxide. Many of the carbonates are well-known mineral rocks such as marble, calcium carbonate, $CaCO_3$.

chemistry – A branch of science that explores the interactions of matter in their various forms. Chemical changes involve alterations in the behavior of the electrons around the nucleus in atoms.

combustion – A term normally associated with the chemical reaction of certain materials with oxygen to produce oxides. The process normally produces heat energy. The combustion of hydrocarbons produces carbon dioxide and water. In cases where oxygen level is low, such as in air, combustion of hydrocarbons can also produce carbon monoxide, CO, a poisonous gas that kills in very small quantities.

compound – Any form of matter made up of two or more *different* elements. For example, water H_2O, carbon dioxide CO_2, table salt, NaCl, etc.

contextual property – a property of matter that does not belong entirely to the object, but depends on external factors. For example, the color of an object depends on a source of illumination. Remove the source of illumination and the color does not exist.

deuterium – an isotope of the element hydrogen having a nucleus containing 1 proton and 1 neutron. The nucleus of a normal (light) hydrogen atom has only 1 proton and no neutrons. The addition of the neutron in the atom deuterium doubles the weight of the atom.

deuterium monoxide – also known as heavy water, it is a water molecule containing at least one deuterium atom in place of a normal hydrogen atom. The properties of heavy water, such as freezing point, boiling point, etc. are much different from normal (light) water H_2O. It is poisonous to humans.

dim wits – by definition, those individuals who believe that greenhouse gas emissions are causing global warming. The term comes about from the observation that the power output of the human brain is only 40 watts.

element – the simplest form of matter that cannot be broken down into simpler forms of matter by any chemical process. To turn one element into another element requires nuclear physics, not chemistry.

electromagnetic radiation – a form of energy involving simultaneous electric and magnetic effects that travels through space at 186,000 miles per second. Generically known as "light," no material object can travel faster than the speed of light.

frequency – as it applies to waves, is the number of times that the medium carrying the wave wiggles up and down in a defined time period. The older unit of frequency was cycles per second. This unit has been replaced by the unit Hertz (after Heinrich Hertz). 1 Hertz is 1 cycle per second. Keep in mind that the wave doesn't wiggle up and down, the medium carrying the wave does the wiggling.

global warming – an evolving term that has come to mean a catastrophic change in climate caused by the burning of fossil fuels. Implications associated with the term are that the globe has never warmed as fast as it has in the last 150 years. It is generally confused with a natural cycle of warming and cooling by the globe over a 200- to 300-year period by persons untrained in science.

greenhouse gas – one of several gases in the earth's atmosphere able to absorb infrared light and transfer the absorbed heat energy to other non-greenhouse gases in the atmosphere.

greenhouse gas effect – an effect in which greenhouse gases, after absorbing heat energy, transfer this heat energy to other gases in the atmosphere thus retarding the escape of this heat energy from the planet into space. The effect is misnamed, since actual greenhouses do not operate this way.

heavy water – see deuterium monoxide.

Henry's Law – a thermodynamic law developed by William Henry in 1803 specifying that the solubility of a gas in a liquid at constant temperature is directly proportional to the pressure of that gas above the liquid.

hydrogen – an element containing 1 proton in its atoms. It is made up of the smallest atoms in the list of elements. Atoms of hydrogen combine to form molecular hydrogen H_2, a dangerously explosive gas.

ice cores – ice obtained by pushing long metal tubes into the polar ice caps and pulling out samples going back thousands of years. Certain regions of Antarctica and Greenland will produce long cores of ice giving a continuous record of the climate at the time the ice was formed.

infrared light – a band of radiation having wavelengths slightly longer (and lower energy) than visible light. This type of light radiation causes atoms in molecules to jiggle and generate heat.

isotopes – two or more atoms of the same element, each with different weights due to a different number of neutrons in the atoms. The isotopes of the element hydrogen are: normal hydrogen – 1 proton, no neutrons; deuterium (heavy hydrogen) – 1 proton, 1 neutron; and tritium (really heavy hydrogen) – 1 proton, 2 neutrons.

Kyoto – a city in Japan where a group of some 37 countries agreed in 1997 to reduce greenhouse gas emissions. The commitment to do this is known as the **Kyoto Protocol.**

light – see *electromagnetic radiation*

Mpemba Effect – the observation, rediscovered by a Tanzanian high school student, Erasto Mpemba, in 1963, that hot water freezes before cold water. While there are a number of theories as to why this sometimes happens, the phenomenon is still not that well understood. A major flaw in the comparison is to assume that the only difference between the two samples of water is their temperature. The effect has been known for ages, mentioned in the writings of Aristotle.

nitrogen – an element containing 7 protons in its atoms. Two atoms of nitrogen join to form a very stable gaseous molecule N_2 at normal temperatures that makes up 78% of the earth's atmosphere.

noise – in science, the random output of a measuring device normally associated with some type of instability in its design. Also, when a series of measurement are made on some constant parameter, the experimental scatter associated with the measurements can be thought of as *noise*. This generally occurs when pushing a measuring device to its limits.

ozone – a molecule of oxygen containing three atoms of oxygen, O_3 with a very characteristic odor. Normally formed in the upper atmosphere, its concentration generally increases after thunderstorms, giving the air a very familiar odor.

oxygen – an element containing 8 protons in its atoms. Two atoms of oxygen join to form a very reactive molecule O_2, a gas at normal temperatures that makes up 21% of the earth's atmosphere. It is an odorless, colorless gas at room temperature. Properties of oxygen gas are not that different from chlorine gas. Oxygen gas, O_2, is a metabolic poison.

periodic – operating in some sort of cyclic fashion, which may or may not be regular.

photosynthesis – a chemical process allowing the compound carbon dioxide and water to react with each other and produce sugars and oxygen. It is the reverse of combustion, and will not take place in nature spontaneously without sunlight and chlorophyll, a substance in plants that absorbs red wavelengths from sunlight and causes the plants to appear green.

physical chemistry – a branch of chemistry that studies the physics of chemical processes.

precision – the relative error between measured values in a series of measurements of the same thing directly related to the measuring device. Sometimes confused with the term *accuracy*. A measurement could be very precise and, at the same time, not very accurate.

pressure – a force pushing on an area of matter used, for gases, to determine the quantity of a gas present in a container of fixed volume and temperature.

salt – a chemical substance resulting from the chemical reaction between and acid and a base. Table salt is sodium chloride, NaCl. Some salts are neutral, like sodium chloride; some salts are basic, like sodium bicarbonate; and some salts are acidic, like ammonium nitrate.

scientific method – the procedure upon which modern science is based that involves the only accepted way that a scientific theory can be validated. Once a scientific theory is proposed, it must be tested in a controlled environment (e.g., a laboratory) to demonstrate that it produces outcomes that are consistent with the theory. If it cannot be tested experimentally, then the theory is not settled science.

universe – as used in this book, an area of study containing a system, which is the focus of the study, surrounded by boundaries, real or hypothetical. Everything outside the boundaries of the system is known as local surroundings.

Vostok – Vostok Station, Antarctica is a Russian research station in the coldest part of Antarctica, opened in 1957 by the Soviet Union, where ice core drilling has produced approximately 420,000 years of ice-core data. The station is still in operation today, a cooperative venture between Russia, the United States, and France.

wavelength – the distance between two equivalent points along a wave. Only a wave oscillating in a regular periodic mode has a well-defined constant wavelength.

LaVergne, TN USA
18 December 2010
209274LV00004B/1/P